W9-DJP-991

Choosing
a Career in
Agriculture

About half of the world's working people are employed in agriculture, far more than in any other industry.

Choosing a Career in Agriculture

Walter Oleksy

The Rosen Publishing Group, Inc.
New York

Published in 2001 by The Rosen Publishing Group, Inc.
29 East 21st Street, New York, NY 10010

First Edition

Library of Congress Cataloging-in-Publication Data

Oleksy, Walter.
    Choosing a career in agriculture / by Walter Oleksy — 1st ed.
      p.   cm. — (The world of work)
Includes bibliographical references.
ISBN 0-8239-3332-6 (lib. bdg.)
1. Agriculture—Vocational guidance—Juvenile literature. [1. Agriculture—
Vocational guidance. 2. Vocational guidance.] I. Title. II. World of work
(New York, N.Y.)
S494.5.A4 O44   2000
630'.203—dc21   00-009603

Manufactured in the United States of America

Contents

The Agricultural Industry

Agriculture is the growing of crops, such as wheat, corn, or potatoes, or the breeding of livestock, such as cows, chickens, or pigs. It is also the most important industry in the world. Our lives depend on agricultural products. Plants and animals provide the food that we eat and the milk that we drink. Other products, such as cotton, wool, and wood from trees, supply basic materials for clothing and shelter. Agricultural by-products also provide materials for other basic needs, such as medicines and paints.

Agriculture is also one of the oldest professions. It began about 11,000 years ago when Middle Eastern tribes discovered how to grow plants from seeds and raise animals for food. Today, about half the working people in the world are employed in agriculture. That is far more than in any other industry. There are three main branches of agriculture: food products, natural fibers, and other agricultural products.

Agriculture is both the growing of crops, such as wheat, corn, or potatoes, and the breeding of livestock, such as cows, chickens, or pigs.

Food Crops

Most food products come from crops that are grown on farms both big and small. The rest of our food comes from livestock, especially cattle and hogs. Fish are also an important part of our food supply, but that's a subject for another book.

The main food crops are cereal grains. They include barley, corn, millet, oats, rice, rye, sorghum, and wheat. Other food crops are root crops, such as potatoes, sweet potatoes, and a tropical plant, cassava. Still other food crops are "pulses"; these include beans and peas. Fruits and vegetables are also important. Farms also grow sugar-yielding crops such as sugarcane and sugar beets. Others grow peanuts, cocoa beans, coffee, or tea.

Other food products come from animals such as cattle, chickens, goats, hogs, and sheep. Livestock supplies nearly all of the world's meat, milk, and eggs. Some farms also raise bees for honey. Others, called fish farms, raise freshwater fish such as catfish and trout, and saltwater shellfish such as oysters and mussels.

Natural Fibers

Some plants are not grown for food. Instead, they provide fibers that have many uses, including raw materials for the clothing we wear. Cotton, flax, hemp, jute, and sisal are among the main plant fibers. Wool, the main animal fiber, comes mostly from sheep but also from goats and members of the camel family, like the alpaca and vicuna. Silk fibers come from the cocoons of silkworms that are raised mainly on farms in Japan and China.

Other Agricultural Products

Some farms do not produce food products or natural fibers. They grow crops to provide raw materials for industry. These materials include natural rubber and animal hides used to make leather. Vegetable oils such as castor oil and linseed oil are used in medicines and paints.

Timber is another very important agricultural product. It is used to build homes and businesses. Most timber comes from natural forests, but there are also tree farms where trees are grown for commercial use. Other farmers raise tobacco crops, or grow flowers, bushes, or other ornamental plants. A few farmers raise foxes and mink for their fur, which is made into coats.

Farming Yesterday and Today

There was a time when a career in agriculture meant backbreaking work, as the farmer got up before dawn and went out to milk the cows, feed the chickens, and plow or harvest the fields. Today, a farmer may spend more time sitting in front of a computer than on a tractor or a milking stool. That's because farming has changed so much. Farming, like just about everything else, has gone high-tech.

You can see this just by looking at the job titles in agriculture. They used to be dairy farmer, pig

Today's farmer will probably spend more time sitting in front of a computer than on a tractor.

farmer, chicken farmer, wheat farmer, and so on. Today, the job titles include biotechnologist, agricultural engineer, agricultural researcher, soil conservationist, irrigation engineer, livestock production manager, and food chemist. They all play their parts in the basic work of agriculture.

There are about two million farms in the United States, but they employ only about 2 percent of American workers. Less than five million people work on farms. There are more unemployed people than there are farmers. The average farm is about 473 acres (191 hectares) and is family owned and operated. However, large commercial farms and other parts of the agriculture industry, such as seed and chemical companies, manufacturers of farm equipment, shippers, and food processors, operate just like large businesses and employ many thousands of people.

Career Opportunities

An estimated 48,000 career opportunities in agriculture are available each year for college graduates majoring in agricultural science. Farm and ranch managers earn more than $36,000 a year. Those working in scientific and technical fields earn $60,000 a year or more. Administrators of large companies that handle farm, ranch, or forest products earn $100,000 a year or more.

You don't have to live in the country or a small town near farms to work in agriculture. Many of the new high-tech jobs are with large corporations in big cities all over the nation and the world. Agriculture is considered one of the high-growth industries of the future. This is because, as the world population expands, more food will be needed. New scientific methods are constantly being sought to produce healthier, more productive plants and animals to feed a hungry world.

Scientists are employed in agriculture as engineers to develop better ways to irrigate crops and design more efficient farm and food production machinery. Chemists develop safe and more effective fertilizers and pesticides to make plants healthier. Genetic engineers modify crop plants and livestock in order to create new or healthier species. Veterinarians look after the health of livestock.

Many careers in agriculture involve education. Farmers need the latest information about new technology and methods of raising abundant and healthy crops and livestock. Jobs are plentiful for vocational teachers of agriculture, adult education teachers, and agricultural college teachers.

Other careers are in communications. Writers, photographers, public relations and advertising people, and radio and television specialists are all needed to report the latest information about the agriculture industry. Weather reporting and forecasting is of vital importance to farmers and others producing crops and livestock.

Where the Jobs Are

Many farms today in the United States, Canada, and other industrialized nations are large commercial farms. There are two main types of commercial farms: specialized farms and mixed farms.

Specialized farms produce one kind of crop or livestock. Most of them rely on advanced technology to produce mass quantities of the product. Many commercial farms raise only one cash crop, such as wheat. In tropical parts of the world, farms producing a single crop such as bananas, rice, coffee, sugarcane, or tea are not called farms but plantations. Growing a single crop is called monoculture.

Livestock raising is highly specialized. There are cattle and sheep ranches, poultry and egg farms, hog farms, and dairy farms. Most ranches ship their cattle to feed lots when the animals are five to twelve months old. The animals are fattened on grain and food supplements and then sent to market.

Poultry and egg farms may be as small as one acre (0.4 hectare), but they may raise 20,000 or more egg-laying hens or broiler chickens. The birds are kept in climate-controlled buildings and fattened with high-energy feed.

Dairy farms raise cows for their milk. They are usually larger than poultry farms but smaller than cattle ranches. Some dairy farmers grow their own feed crops, but others buy feed from commercial suppliers.

Mixed farms also are called diversified farms. A variety of crops are grown and livestock is raised as well. This type of farm has existed since colonial times, but most farms today are highly specialized.

You'll meet some farmers in the next chapter. They will tell us what work is like on crop and dairy farms.

Working on a Farm

Farming has changed a great deal over the past fifty years. Small family farms are not as plentiful as they once were. For a variety of reasons, mainly economic, over 35,000 farms close each year in the United States. The farms that remain become larger and produce more.

Farm Work

Working on a farm is not easy. It requires strength and the willingness to work outdoors in all kinds of weather. The hours are often long, from sunup to past sundown. The work follows an annual cycle to coincide with the growing and harvesting of crops.

In the spring, farmers plow the fields, put down fertilizer, and plant seeds. In the time that's left, they work at repairing farm buildings and machinery. Then, in the summer, the fields need hoeing and weeding, and some crops are harvested. More repair work on equipment or the painting of farm buildings takes up any spare time. In the autumn, the main harvesting is done and the fields are

In the springtime, farmers plow fields, put down fertilizer, plant seeds, and repair farm buildings and machinery.

cleared for next spring's planting. Machinery and buildings are prepared for winter. In the winter, farmers tend to do more repair work, fix fences, and get everything ready for spring planting.

If a farmer raises animals, they have to be fed and given water daily. Cows need to be milked two or three times a day. If a farmer and his or her family, including children of work age, cannot do all the work, others are hired to help. Extra help is usually needed during spring planting and fall harvesting.

It helps to be athletic and healthy. Farm workers have to be able to stand, crouch, kneel, and move around easily. They may have to lift up to 100 pounds. They have to tolerate dust, heat, cold, rain, snow, and long working hours.

The jobs available on large farms include farm manager/operator, field crop manager, livestock farmer, breeder, farm worker, dairy laborer, and dairy herdsperson. Today's successful farmer must keep up with changes and improvements in the farming industry to remain competitive.

Large Corporate Farms

The largest farms, about 3 percent of all farms, are usually owned by large corporations. Many people are hired to work on these farms. One of the positions a person might have is that of farm manager. The farm manager is in charge of farm operations. This may mean supervising the entire operation or just one part of it. The manager may work for a farm management company and be assigned to manage one farm or several farms. A farm manager supervises the other workers. He or she also supervises the farm's production and handles all financial matters.

Another position on a corporate farm is the farm operator. Farm operators either own their own farm or are "tenants" who work the land owned by other farmers who rent the land to them. They may also supervise the work of others on the farm. On very large farms, there may be over 100 workers, including truck drivers, sales representatives, computer specialists, and bookkeepers.

Crop Farms

Some farms raise food grains, such as wheat and rice, for human consumption. Others raise feed grains, such as corn and sorghum, that are fed to poultry and cattle. There are many different careers in crop production. Agricultural engineers design

Farming can be highly specialized; some farmers concentrate exclusively on growing fruit, vegetables, rice, cotton, peanuts, or seeds.

farm equipment, buildings, and farm water systems. Agronomists, who know about soil and crop science, make plants grow healthier and more abundantly. Entomologists and plant pathologists help to protect crops and farm animals from insects and diseases.

Farmers themselves can be highly specialized. They can concentrate on growing fruit, vegetables, rice, cotton, peanuts, or even seeds for other farmers. Of all the careers in agriculture, working on or owning a small farm is one of the riskiest businesses today, as the following interview explains.

Kris, an Iowa Farmer's Daughter

Kris, who grew up on her parents' farm in Iowa, worries about the future of the American family farm.

My grandparents were Iowa farmers. Since I was a little girl, I worked on my parents' grain and hog farm of about 500 acres.

But small farms like ours are a dying breed. Just fifty years ago there were more than 182,000 small farms in the state. Now there are only about half that number. The others went out of business because of low market prices and competition from big corporate farms and foreign countries that export farm produce at lower prices because their labor is cheaper than in this country.

Farming doesn't bring in what it used to. Dad used to sell our corn for $2.75 a bushel, but today it's only worth about $1.93. Hogs used to sell for 55 cents a pound, but last year dad only got a dime a pound. Meanwhile, the costs of farming keep going up. A tractor that used to sell for $25,000 now costs over $75,000. A combine that sold for $40,000 now costs $200,000.

I wanted to stay and keep working on the family farm, but it couldn't support my folks and me. So I went to college and then got a job in Chicago. I'd much rather be back on the family farm. If it's still there tomorrow.

It's sad when small family farms like ours die. When they die, whole towns

*like the one I lived in die, too. A whole
way of life dies.*

*But my folks are stubborn and
tough. They've lived through lots of hard
times. They'll keep farming their 500
acres no matter what. Now all the pigs
are gone. Mom just applied for a job off
the farm. She says she'll help to save the
farm for her grandchildren.*

Livestock and Dairy Farms

Most Americans and Europeans get almost half of
their calories from eating meat products. This
comes mainly from beef cattle and dairy cows that
produce meat and milk. Careers at livestock and
dairy farms include cattle ranching, dairy farming,
dairy herding, livestock production, and animal
breeding. Many jobs on these farms involve a lot
more than just milking cows. They require science
backgrounds in many aspects of livestock care,
such as diet, health, and reproduction. Farmers
need the services of bacteriologists, dairy food
chemists, microbiologists, and sanitation experts.

Dave, Wisconsin Dairy Farmer

Dave, who works and lives with his wife and three
children on a 900-acre, 200-cow dairy farm in
Wisconsin, tells about the life of a dairy farmer. The
farm is owned by another family.

*I had a dairy farm of my own a few
years ago. But I lost it when I couldn't*

get a loan to keep it going. There was just too much competition from large dairy farms for a bank to finance me. So I gave up the homestead and got a job helping a nice family with their bigger dairy farm.

I'm really a jack-of-all-trades—herds-man, breeder, resident mechanic. I run a tractor and till the soil for feed crops. I always get up before sunup to milk the cows at 6 AM. Then I do all the other things before milking again at 2 PM and 9 PM. Each milking takes over two hours.

I've been milking cows since I was in high school—by hand back then, but it's all mechanized now. We have a rotary milking parlor. The cows line up in a circle in stalls and are milked automatically. It's a scientifically designed setup where the cows go through a sprinkler pen, then a drip pen, before milking.

My wife and kids and I love the dairy farm life. It's healthy living in the country. My eldest son is nineteen and just finished a sixteen-week college course in agri-cultural science. He can't wait to have his own dairy farm. I hope he makes it.

Farming is hard work and the hours are long, but it's my choice. The pay is good. I make about $35,000 a year. My wife helps financially by working for a business in a nearby town. It's not easy being a dairy farmer, but it's still a good life.

Milking cows has become a highly mechanized process. Machines such as the rotary milking machine can milk cows automatically.

Questions to Ask Yourself

There are advantages and disadvantages to working on a farm. Would you be happy living in the quiet country away from a busy city? Do you want to be your own boss, or do you want someone else to tell you what your work will be? Are you willing to work long hours outdoors in hot or cold weather and rain and snow? Are you willing to learn about new technology to make your work more productive?

Agricultural Science

B esides working on farms, there are many other career opportunities in vocations that support agriculture. Many of these are in the various agricultural sciences. Private industry, universities, and the government employ thousands of agricultural specialists who work in the scientific side of agriculture.

These men and women are trained in animal and crop science, horticultural science, and environmental assessment. They also monitor land use, soil conservation, and food processing, and they make sure that government regulations are followed in the growing and preparation of food. Others conduct laboratory research to create healthier and more abundant crops and livestock.

Many agricultural scientists earn as much as $35,000 to $75,000 a year or more. Banks hire them as advisors on farm management and use them to help farmers succeed. The feed industry relies heavily on the expertise of agricultural scientists to ensure that animal feed is healthy and nutritious.

Agricultural scientists who work in soil management conduct research to find ways to make soil richer and healthier so that crops can be grown in all climates.

Soil Science

Some agricultural scientists work in the very important field of soil management. They conduct research to find ways to make the soil richer so more and healthier crops can be grown to feed the world. Pedro, a scientist who specializes in soil, tells of his work for a university in the United States that takes him to the farms and rain forests of Central and South America. There he helps local farmers discover new methods of soil management that will increase crop production.

An ancient method of soil management is "slash-and-burn" farming. Areas of forest are cut down and burned. The ash fertilizes one or two seasons of crops, but then the land loses its nutrients. Farmers then have to leave the area and do the same thing somewhere else. Pedro explains why that is a bad idea for farmers.

Soil conservationists consider slash-and-burn an unproductive method of farming. It also contributes to tropical deforestation, a major cause of global warming.

I became an assistant professor of soil management at North Carolina State University. They assigned me to study soil conditions in various parts of the world. For the past few years, I've been head of a soil management project in the jungles of Peru.

Working conditions were crude in the beginning, but they've improved. The villagers have been wonderful, eager to get help in learning new and better ways to grow their crops.

I feel my years at the jungle research station in Peru are the highlight of my career. I grew up on a farm in Cuba. Now I'm helping find ways to improve soil and crop management for a world growing rapidly in population. Good soil management helps produce more and healthier food, especially in those countries where drought or war has destroyed crops and is causing widespread famine.

Soil science is a vital career in agriculture. It can be an especially good

*field for young people who want to work
in other parts of the world.*

Biotechnology

Another very important career in agriculture is in biotechnology. This involves the development of chemicals that improve plant growth or protect plants from predators, as well as the genetic modification of plants to improve their usefulness to people. Another goal of bio-technology is to create renewable sources of energy from plants.

"Biotechnology is becoming a key resource for many industries working in pharmaceuticals, agriculture, chemical products, energy conser-vation and new energy sources, and solutions to environmental problems," says Daniel D. Godfrey, dean of the North Carolina A & T State University School of Agriculture.

Upon graduation, students enrolled in a biotechnology certification program receive both an undergraduate degree and a certificate in biotechnology. Coursework for the certificate includes studies in animal science, biology, chemical engineering, agricultural chemistry, natural resources management, and environ-mental design.

For more information on biotechnology careers, contact the Biotechnology Industry Organization. Its address and telephone number can be found in the For More Information section at the back of this book.

Genetic engineering of plants and animals, like the pigs shown here, is a controversial subject and one that raises many ethical issues.

Genetic Engineering

Genetically altering plants to produce more and healthier species is one of the most interesting and challenging careers in agriculture today.

Just a few years ago, scientists cloned a sheep named Dolly. Cloning creates an exact genetic duplicate of a plant or animal in the laboratory. Cloning is important because it can increase production of plants and animals and perhaps create healthier, more nutritious copies of organisms than exist in nature. However, it is a highly controversial technique and many people are opposed to it for religious or ethical reasons.

As this book was being written, progress was reported in the effort to clone cows. It came from researchers at Advanced Cell Technology, a company that uses cloning to treat human disease and extend life. The news was that the cows produced in cloning experiments had youthful cells and should live at least as long as ordinary cows. This was important because tests on Dolly the sheep revealed that the tips of her chromosomes were unusually short and that she might not live as long as regular sheep.

However, biotechnology and genetic engineering as applied to food plants and livestock is a controversial subject. Some critics of the new technologies are concerned that gene manipulation as applied to plants and livestock may not be safe. Advocates of the new technologies say they are safe and that crops have been engineered to stay fresher longer and to resist pests and herbicides without damaging the soil, air, or water.

Addressing concerns about biotechnology and genetic engineering in agriculture in May 2000, President Clinton's administration announced a new effort to assure the public about the safety of biotech foods. The plan included a process for reviewing genetically altered crops, as well as new standards for food makers to follow in labeling products. The plan would require biotech companies to notify the Food and Drug Administration (FDA) at least four months in advance of releasing new genetically engineered organisms for food and animal feed. It also would require biotechnology companies to provide the FDA with their research data.

Agricultural Chemistry

Agricultural chemistry is not a distinct discipline. It ties together genetics, physiology, microbiology, entomology, and other sciences. Agricultural chemists help to create more productive plant and animal strains. They determine the kinds of nutrients that are needed for the healthy growth of plants and animals. They also determine a soil's ability to provide essential nutrients for the support of crops for human consumption or for livestock feed.

Each year in the United States, millions of pounds of agricultural chemicals are used in fertilizers, herbicides, fungicides, and insecticides that are put on crops. One of those working as a chemist and environmental studies coordinator for the agricultural products division of a major chemical company tells about her job.

Julie, Agricultural Chemist

I help make sure that the chemicals used in farming are safe for livestock, humans, and the environment. These chemicals are applied in huge quantities to commercial crops. In a test field, we study the effect of these chemicals on the air, plants, soil, and water. A large presence of these chemicals may pose a threat to humans and animals, or may pollute the air and water.

I work with a lot of scientists in other fields, including agronomists, biologists, and biochemists. My research tells them the toxic level of a product in the field, and they tell me its impact on animals and plants.

There is a wide range of research projects in agricultural chemistry. So it's important to be able to work in other fields, or at least to know something about them. It's not good to have tunnel vision. You need to be as well-rounded as possible.

Ed, Agricultural Chemist

Similar work is done by Ed, a chemical specialist who works with the United States Department of Agriculture.

I focus on the journey of pesticides, discovering the unexpected places where they may eventually turn up. I test for agricultural chemicals in the atmosphere, surface water, and groundwater. In some cases, the research I do results in changes in the way that some chemicals are used.

Agricultural chemists usually work in a laboratory or at the site of a field test. They are employed by government agencies such as the U.S. Food and Drug Administration and the Environmental Protection Agency. Some of the larger

Agricultural chemists usually work in a laboratory or at the site of a field test and are often employed by government agencies.

chemical companies consider their agricultural divisions to be their most lucrative businesses, and employ many chemists. Agricultural chemists also find work as teachers in agricultural schools.

Agricultural chemists need to know about a wide range of farm matters, such as crops, weeds, soil, air, and animals. Good communication skills are important since we often work with others as part of a team in studying potential environmental concerns.

It's very helpful to take courses in biology, biochemistry, human toxicology, water and soil chemistry, and geology. Also, it's good to know how to use computers.

Starting salaries for those with bachelor's or master's degrees in agricultural chemistry range from about $25,000 to $35,000. Those with doctoral degrees may start at salaries as high as $50,000. Agricultural chemists in government jobs usually earn a little less.

For more information about careers in agricultural chemistry, contact the American Crop Protection Association. Its address and telephone number can be found in the For More Information section at the back of this book.

Agricultural Engineering

Agricultural engineers are needed in food and livestock production to design new farm and factory buildings and farm equipment. They also design systems to ensure that a farm or breeding facility is environmentally safe. They are working on some very far-out ideas; engineers are currently working on developing driverless tractors that would be guided by either preprogrammed computers or radar. While that particular achievement may be a long way off, agricultural engineers do keep discovering new and more realistic ways to help food production and distribution. Agricultural engineers earn from $50,000 to $75,000 a year or more.

Career Planning

If you're considering a career as an agricultural engineer, you should first learn the basics of engineering. Colleges and universities that teach agricultural engineering will offer courses that include mathematics, chemistry, theoretical and applied mechanics, hydrology, hydraulics, electricity, and magnetism. Students learn about forces, materials, power applications, machine design, and other aspects of general engineering.

After learning general engineering, a branch of engineering such as agricultural engineering can be considered. There are several branches of study to choose from within the field of agricultural engineering. These include farm power and machinery, farm structures, and soil and water control.

Most agricultural engineers are interested in farm power and machinery and work in that specialization. They design and test experimental machines and prototypes for large farm machinery companies. They also work for companies that provide the parts for the hi-tech machines. Other agricultural engineers design petroleum products for gas and oil companies. Still others work for manufacturers of crop-drying equipment.

Agricultural engineers specializing in electric power work with electric power suppliers, manufacturers of electrical equipment, and other companies that supply the agriculture industry. Those working on creating better farm structures design buildings that can efficiently store grains or livestock feeds. Designing structures that can improve a farm's productivity is important to farmers because more than one-fifth of their total capital outlay is invested in their farm buildings.

Most of the work of agricultural engineers in soil and water control is concerned with keeping

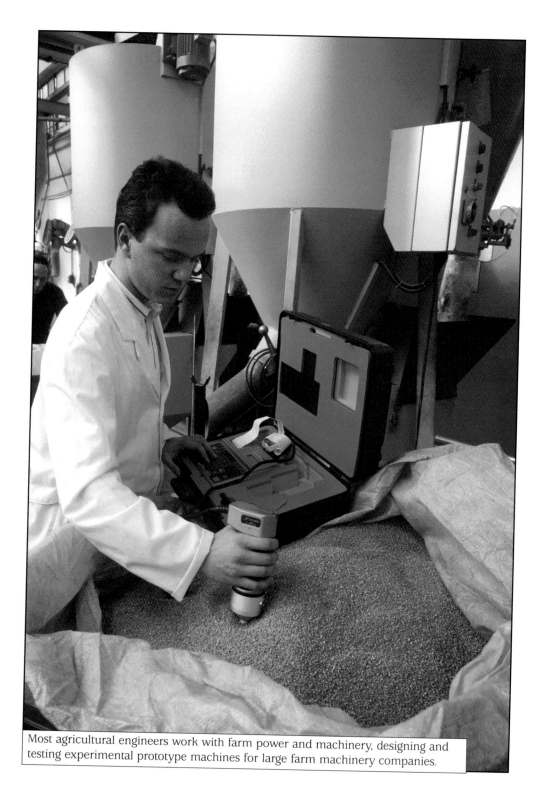

Most agricultural engineers work with farm power and machinery, designing and testing experimental prototype machines for large farm machinery companies.

farmland nutritionally healthy, minimizing soil erosion, and keeping irrigation systems working efficiently. Farm water systems are often at risk from the construction of nearby highways and airfields. In these areas of specialization, agricultural engineers often work in teams. Team specialists may include civil engineers, soil physicists, agronomists, and other specially trained scientists.

While an undergraduate degree in agricultural engineering is recommended, a master of science degree or a doctorate in the field assures an even more successful career. More and more, engineers are combining computers and electronics in their design of machinery for agricultural uses. On large dairy farms, for example, a computerized sensing device is attached to a cow's ear. It records how much milk the cow produces and adjusts its ration of feed accordingly.

Other agricultural engineers are involved in solving irrigation problems for farmers. For example, engineers recently devised a more efficient method of irrigating citrus orchards in California. Some environmental engineers study land erosion problems to reduce soil loss. Others develop waste treatment systems for vegetable and chicken processing factories, or solve problems of storage and disposal of dairy and livestock wastes.

Agricultural Research

New and more productive ways of growing crops and raising livestock are constantly improving agricultural methods around the world. A wide variety of careers in agricultural research are available at private companies, government agencies, and universities.

Careers in agricultural research cover a wide range of fields. For the most part, they involve studying crop and livestock production and how harvests are processed for the marketplace. Agricultural researchers are highly educated and trained, and work with the latest high-tech equipment.

One of the leaders in agricultural research is the Agricultural Research Service (ARS) of the U.S. Department of Agriculture. The ARS's primary mission is to study ways to conserve and wisely manage soil, water, and air resources. The ARS has facilities in 127 locations in the continental United States and in Alaska, Hawaii, and Puerto Rico. These facilities are located in every major farm and rangeland ecosystem, where about 8,100 ARS employees work on various agricultural problems.

Agricultural researchers work with the latest high-tech equipment to find more productive ways of growing crops.

About one-third of the researchers are agricultural scientists and engineers. The rest provide technical support. The ARS employs biotechnologists, soil scientists and conservationists, insect specialists, and geologists.

Agricultural researchers also earn good salaries, like those in agricultural science and engineering. Salaries range from $35,000 to $75,000 or more a year.

Maria, Entomologist Researcher

I never thought of agriculture as research. My family were migrant farmers in Mexico and then Texas. To us, agriculture was going out to work in the fields, picking melons and onions.

All that changed when I went to work for the ARS between my junior and senior high school years. I became a research apprentice in the Honey Bee Research Unit of the Subtropical Agricultural Research Laboratory at Weslaco, Texas.

Two of my sisters soon followed me in working at different research stations for the ARS. Eventually, we all got college degrees in science or engineering. I'm working on my doctorate in entomology and I just accepted a job researching integrated pest management at Texas A&M University's agricultural experiment station in Dallas.

The ARS research apprenticeship program gave us a fantastic start toward successful careers. It was right at our own back door.

The ARS has a wide range of beginning career opportunities for young people, according to Floyd P. Horn, administrator of the ARS. "Whether collecting field data, dissecting insects, preparing tissue cultures, or purifying nucleic acids, students get firsthand exposure to research. It helps young people cultivate a scientific mind-set and a variety of research skills." Victor Commisso, personnel management specialist with the ARS Recruitment and Employment Office, advises, "The best time to seek a federal career in agricultural research is while you're a high school student. If you're serious about it, the time to pursue it is now."

Both temporary and permanent career opportunities are open for students who are at least sixteen years old and studying at least half-time at a high school, technical or vocational school, or university. Those interested should send Victor Commisso's office a résumé and cover letter telling about their career goals and interest in a student position.

For more information on the ARS student programs, contact the Human Resources Division of the U.S. Department of Agriculture Research, Education, and Economics. Its address can be found in the For More Information section at the back of this book.

Environmental Protection and Agriculture

A lmost six billion people live on Earth. In order to make sure that everyone today and in future generations is fed adequately, soil, water, air, and sunlight have to be used wisely. It is the job of agricultural environmentalists to work toward that goal. Through the application of science, they try to protect and conserve the planet's natural resources for present and future food needs.

Increasingly, computer technology is incorporated into the work of researchers in environmental engineering. An example is the computerized Geographic Information System (GIS) that was developed by agricultural engineers at Michigan State University. Information taken from individual farms by geologists, soil scientists, pesticide scientists, and others is analyzed by the GIS, which then reports on areas of potential groundwater contamination. Salaries for those in environmental work related to agriculture range from $35,000 to $75,000 or more a year.

Computer technology is used by environmental engineers to analyze soil and groundwater for pesticides and other contaminants.

Traditionally, engineering—including agricultural engineering—was more of a man's occupation than a woman's. That is changing, however, as the next interview with a female Hispanic-American agricultural engineer proves.

Elena, Environmental Engineer

I became interested in engineering because a boyfriend was studying it in high school. I got a degree in civil engineering and decided that the work I liked best was the part engineering plays in the environment.

I began my career with a major oil company. For over ten years, I have worked all over the country on environmental projects. One of them was helping to make the company's equipment for turning regular gasoline into unleaded gasoline safe for the environment. This required an understanding of chemistry and the effects of temperature and high pressures.

On another environmental program, I supervised the replacement of worn underground steel gas tanks at service stations with fiberglass tanks. The steel tanks had to be replaced before they started leaking. I worked with computer models of the steel tanks that predicted their life span, which depended on their age and the soil they were in. Doing the job required a knowledge of statistics and geology in order to determine which tanks should be replaced first.

Environmental engineering is a great career and it's wide open for women. I especially encourage Hispanic-American women to follow me into this very important field.

Other Careers in Agriculture

There are many other career opportunities in agriculture. They involve the economics of agriculture, selling what farmers produce, and teaching others about new agricultural developments. Other careers are in the fish and timber and forestry industries. Still others are in reporting on agriculture in the print media and for radio and television. Salaries in these careers range widely, from $25,000 to $75,000 or more a year.

Economics of Agriculture

Today, successful farming of any kind depends on successful financial management. The U.S. Department of Agriculture offers career opportunities in its Economics Research Service (ERS) to those who want to help farmers in this aspect of their work. Those who work for the ERS analyze and forecast agricultural production and demand, not only in this country but worldwide.

People who work for the ERS evaluate how successful farms are in producing and marketing

Other careers in the field of agriculture include economic research, forest service, agricultural teaching, and retail.

what they grow or raise. They estimate the effects that government regulations and programs have on both farmers and consumers. A college education in economics and statistical analysis is essential for this kind of work.

Forest Service Careers

An important part of agriculture is the growth of new trees and the management of timber farms and state and federal forests. The U.S. Department of Agriculture has eight regional Forest Service centers. At these centers, researchers and forest management specialists work on problems dealing with forestry. One of these is the dispute between logging companies that want to cut down more trees in national forests and environmentalists who want to save them.

Careers in forestry also involve forest inventory, fire fighting, insect and disease study and control, watershed management, fire management, wildlife and fish habitat management, and the management of public recreation in parks and forests.

Careers in Agricultural Education

Many young people become interested in agricultural careers while in high school. Over 7,000 high schools offer agricultural career courses in cooperation with the Future Farmers of America. The programs are funded in part by the National Vocational Education Act. The funds help provide salaries for about 12,000 teachers of agricultural science in high schools across the United States.

State and federal agricultural programs are also in need of teachers. Many of these are called "cooperative agricultural extension service" programs. That is because of the programs' joint funding by county, state, and federal governments. Careers in the agricultural extension service range from county extension agents to administrators of various farm programs.

More than 25,000 young men and women graduate each year from agricultural colleges. The schools are always in need of qualified teachers of agricultural subjects, as well as those with backgrounds in math and science.

Agricultural Journalism

Careers are numerous in writing and reporting about agriculture. Journalism opportunities include writing about farms and agricultural issues in newspapers and magazines. Many magazines specialize in reporting on a specific industry, such as dairy or grain production, or more general subjects, such as nutrition.

Careers in agricultural journalism also include radio and television reporting of news of interest to those working in agriculture. While local radio and television stations are hiring fewer people to broadcast farm reporting, the larger networks are increasing their hiring. Broadcast networks provide the essential information to local radio and television owners who can't afford to hire a staff broadcaster of agricultural news.

Getting a Job in Agriculture

Today, a solid education in agricultural science is necessary to succeed in a career in agriculture. Every farmer must be knowledgeable about the science behind crop production, livestock management, soil and water conservation, and all other aspects of agriculture.

High School

Many high schools in every state offer instruction for careers in agriculture. About 640,000 students attend these courses each year. Many high school agriculture courses include both classroom study and practical field experience. Students may be required to raise a crop or a farm animal or work on a farm or in an agricultural business.

Colleges and Universities

Today, a college degree in agricultural science is essential for a successful career in agriculture. About 21,000 students receive bachelor's degrees

Today, a career in agriculture requires a solid education in agricultural science, including crop production, livestock management, and water and soil conservation.

in agriculture each year from colleges and universities. About 6,000 others receive master's or doctorate degrees in some field of agricultural science. While in the past many of those who enrolled in an agricultural science degree program were from farm families, today's students come from a broader range of backgrounds.

"It used to be that many of our students came from farms and intended to return to farming after graduation," says Daniel D. Godfrey, dean of the North Carolina A&T State University School of Agriculture. "Now, young people who might not have even visited a farm, but who know the importance of natural resources, are coming to us. They are aspiring toward highly technical and potentially lucrative careers in agriculture and the natural sciences."

Like many other universities, North Carolina offers studies in biotechnology, agricultural engineering, natural resources management, waste management, and food microbiology. "There continues to be a great demand for people skilled in the application of science and technology to food and natural resources," says Godfrey.

The School of Agriculture recently conducted a live video-conference on careers in science and technology. A recording of the conference can be seen on the Internet at *http://www.ag.ncat.edu/video/Online.htm*.

Youth Programs

If you're interested in a career in agriculture, you can learn how to get started in high school. Ask your school's science teacher or career counselor about it. Knowledge and interest in biology and general science is important, as are skills in math, English, and accounting. In addition to schools where young people can learn more about careers in agriculture, there are two major youth programs. They are the 4-H Clubs and the Future Farmers of America. Both organizations provide information and counseling on agricultural careers. You don't have to live in a small town near a farm to join either organization.

4-H Clubs

The nationwide 4-H Clubs were founded in 1905 to help boys and girls learn more about careers in agriculture. These clubs also help young people to develop character skills through shared interests and activities in farming. The 4-H Clubs' motto reflects these goals. The four H's stand for:

Head: Clearer Thinking

Heart: Greater Loyalty

Hands: Larger Service

Health: Better Living

4-H Clubs began nearly a century ago when people in rural communities saw that many children of farm families were turning away from farm careers and moving to the cities for jobs in business and industry. The clubs continue to encourage young people to pursue careers in agriculture. They also offer guidance in learning the high technology aspects of agriculture. At present, about five million boys and girls are members of 4-H Clubs throughout the United States.

Future Farmers of America

Future Farmers of America was organized in 1928 to help young people learn about and prepare for careers on farms and in other branches of agriculture. About 430,000 young people are members of the FFA. The organization helps them prepare for successful careers in all fields of agriculture, from farming to mastering all the new technologies associated with food and livestock production and management. The FFA also works to develop young people's potential for leadership, personal growth, and career success through agricultural education. Both boys and girls can join local FFA chapters that are organized at their schools.

The Importance of a College Degree

Even to work on a small farm, a college degree in agricultural science can mean the difference between success or failure. Many colleges offer bachelor of agricultural science degrees. Some require four years of study, others six years. The programs stress that agricultural science is not only about becoming a farmer. The degree provides thorough training for professional scientists, with a minor in economics. This is because modern agriculture is a very competitive business.

Tech Support for Farmers

Only a small number of those with bachelor of agricultural science degrees become farmers, or in today's terms, land managers. But today's farmers need the support of skilled technicians in a variety of fields, from the design of machinery to the chemistry of food processing, and many who study agricultural science find careers in these related fields.

Agricultural science courses provide a basic understanding of the relationship between plants, soil, the environment, and people. The emphasis is

Agricultural science is the study of the relationship between plants, soil, the environment, and people, emphasizing science and economics.

on science and economics. Students not only learn about modern methods of farming and agricultural science in the classroom—they take part in on-the-job research projects. They also are assigned to work programs on farms or in agricultural industries. This leads to an understanding of the whole rural environment.

The Future of Farming

Farming is undergoing great changes today. Small farms are disappearing by the thousands every year, and the total number of people working on those farms is also declining. The remaining farms are growing larger and are being managed more and more like industries. But for that very reason, today's farms require the services of more skilled specialists. Modern farming has become a knowledge-based industry, more dependent on science and less dependent on the experience and wisdom of the traditional farm family working the land. This has opened up new opportunities for people interested in agriculture who do not have a traditional rural background. However farming evolves, you can become a part of it and know that you are contributing to a stronger, healthier nation.

Glossary

acre   A parcel of land; an acre of land is about the size of a soccer field.

agricultural chemists   Those who study the application of chemicals in farming.

agricultural engineers   Engineers who design farm equipment, buildings, and farm water systems.

agronomists   Those who study the raising of crops and the care of the soil.

biotechnology   The creation of new products and improved food plants using the techniques of modern biology.

cash crops   Crops sold for money instead of being consumed on the farm.

cloning   Artificially producing an exact copy of a plant or animal.

corporate farms   Large farms owned by companies rather than individual people.

dairy farms   Farms on which cows are raised for their milk.

diversified farms   Farms on which several crops are grown and livestock is raised; also called mixed farms.

economists   Those who study the production, distribution, and consumption of goods and services.

entomologists   Scientists who study ways to protect crops and livestock against insects and diseases; also called plant pathologists.

environment   The soil, water, air, and other natural qualities of an area such as a farm.

feed   Grain and other food fed to farm animals.

fertilizer   Chemicals added to soil to help plants grow better food crops, mainly cereal grains such as corn and wheat.

genetic engineering   In agriculture, altering the genes of plants and livestock to produce new strains.

harvest   To gather crops for market.

hydraulics   The science dealing with liquids in motion.

hydrology   In agriculture, the science of studying the condition and circulation of water on a farm.

irrigation   A system designed to supply farms with water by artificial means.

livestock   Farm animals such as cows, chickens, and hogs.

natural fibers   Plants grown for use in making fabrics, including clothing.

plantations   Large farms growing a single crop, such as bananas, rice, coffee, sugarcane, or tea.

pulses   Mainly dry beans and dry peas.

soybeans   Type of bean for making oils, flour, and other foods.

specialized farms   Farms producing mainly one kind of crop or livestock.

tenant farmers   Hired people who farm land for others who own it.

For More Information

In the United States

Agricultural Research Service
U.S. Department of Agriculture
5601 Sunnyside Ave.
Beltsville, MD 20705
Web site: http://www.ars.usda.gov

American Crop Protection Association
1156 15th Street NW, Suite 400
Washington, DC 20005-1716
(202) 296-1585
Web site: http://www.acpa.org

Biotechnology Industry Organization
1625 K Street NW, Suite 1100
Washington, DC 20006-1621
(202) 857-0244
Web site: http://www.bio.org

Defenders of Wildlife
1101 14th Street NW, Suite 1400
Washington, DC 20005
(202) 682-9400
Web site: http://www.defenders.org

Food and Agricultural Careers for Tomorrow
Purdue University
1140 Agricultural Administration Building
West Lafayette, IN 47907-1140
(765) 494-4600
Web site: http://www.agriculture.purdue.edu

Future Farmers of America
National FFA Organization
6060 FFA Drive
P.O. Box 68960
Indianapolis, IN 46268-0960
(317) 802-6060
Web site: http://www.ffa.org

Greenpeace USA
702 H Street NW
Washington, DC 20009
(800) 326-0959
Web site: http://www.greenpeaceusa.org

National 4-H Club Headquarters
1400 Independence Avenue SW
Washington, DC 20250-2225
(202) 720-2908
Web site: http://www.4h-usa.org

In Canada

Canada Agriculture Online
http://www.agcanada.com/cao.htm

The Canadian Farmer
http://www.canadianfarmer.com

Canadian Rural Partnership
http://www.rural.gc.ca

For Further Reading

Clark, Robert, ed. *Our Sustainable Table.* San Francisco: North Point Press, 1990.

Fish, Charles. *In Good Hands: The Keeping of a Family Farm.* New York: Farrar, Straus & Giroux, 1995.

Kallen, Stuart. *The Farm.* Minneapolis: Abdo & Daughters, 1997.

Kunhardt, Edith. *I Want to Be a Farmer.* New York: Grosset & Dunlap, 1989.

Oleksy, Walter. *Hispanic-American Scientists.* New York City: Facts on File, 1998.

Paladino, Catherine. *One Good Apple: Growing Our Food for the Sake of the Earth.* Boston: Houghton Mifflin, 1999.

Roop, Peter, and Connie Roop. *A Farm Album.* Des Plaines, IL: Heinemann Library, 1999.

White, William C., and Donald N. Collins. *Opportunities in Farming and Agricultural Careers.* Lincolnwood, IL: VGM Career Horizons, 1996.

Index

About the Author

Walter Oleksy is the author of over forty books for young readers. His other books for Rosen include career books on fire fighting, video game design, and Web page design, as well as books on the nervous system and the circulatory system. He also has written biographies of Christopher Reeve, Princess Diana, and James Dean. You can read more about Oleksy's books at *http://home.earthlink.net/ ~ waltmax/bio.html*. Oleksy lives in a Chicago suburb where he loves to work in his flower and vegetable garden and take his dog, Max, for long walks in the nearby woods.

Photo Credits

Cover © Michael Melford/The Image Bank; p. 2 © Richard Gaul/FPG; pp. 7, 30 © Telegraph Colour Library/FPG; p. 10 © Westerman/International Stock; p. 15 © Richard Gaul/FPG; p. 17 © J.G. Mason/International Stock Photo; p. 21 © Anthony Cooper/Ecoscene/CORBIS; p. 23 © Chris Salvo/FPG; p. 26 © Reuters Newmedia Inc./CORBIS; p. 34 © Royalty Free/CORBIS; p. 37 © Michael Lichter/International Stock Photo; p. 41 © Ed Young/CORBIS; p. 45 © Andre Jenny/International Stock; p. 49 © John Zoiner/International Stock; p. 53 © James P. Blair/CORBIS.

Series Design

Geri Giordano

Layout

Danielle Goldblatt